Ekaterina Samigulina

Modification of all types of fuel mixtures

Ekaterina Samigulina

Modification of all types of fuel mixtures

Complex innovation projects as a consequence of remote control and monitoring application

ScienciaScripts

Imprint
Any brand names and product names mentioned in this book are subject to trademark, brand or patent protection and are trademarks or registered trademarks of their respective holders. The use of brand names, product names, common names, trade names, product descriptions etc. even without a particular marking in this work is in no way to be construed to mean that such names may be regarded as unrestricted in respect of trademark and brand protection legislation and could thus be used by anyone.

Cover image: www.ingimage.com

This book is a translation from the original published under ISBN 978-620-8-17026-4.

Publisher:
Sciencia Scripts
is a trademark of
Dodo Books Indian Ocean Ltd. and OmniScriptum S.R.L publishing group

120 High Road, East Finchley, London, N2 9ED, United Kingdom
Str. Armeneasca 28/1, office 1, Chisinau MD-2012, Republic of Moldova, Europe
Printed at: see last page
ISBN: 978-620-8-22652-7

Ekaterina Samigulina

Modification of all types of fuel mixtures

Integrated innovation projects as a consequence of remote control and monitoring applications

Keywords:

Dynamic homogenisation, Complex modification of all fuels, Design advantages, Emulsion formation in the flow of its components, Radial channels, Turbulence homogenisation, Coaxial conical ring channels, Exhaust gas recirculation system, Bernoulli effect, Range of technological possibilities.

Annotation

As the statistics of innovation projects shows, the algorithmic component, including the logistics of the entire innovation process, starting from the formulation and synthesis of an innovative idea and ending with the process of integration into specific production and commercial structures, has an increasing impact on the commercial value and efficiency of these projects.

The setting of the objective, the choice of evaluation criteria and the nature of the pathways to achieve the required results often determine the success or failure of the implementation process.

Since the author has experience and technological and commercial developments in the most demanded today technologies of fuel mixture modification, he proposes as an example to consider the algorithmisation of this group of innovative projects, which are the final stage after passing the multifunctional stages of transportation and delivery, at that, as a rule, including system remote monitoring, including combined with the processes of aerial photography or with the use of drones.

TABLE OF CONTENTS

INTRODUCTION

An original, repeatedly tested technology has been created for complex modification of all types of fuel mixtures.

The basic benefits of technology.

Design advantages of the device for dynamic formation of emulsion in the flow of its components. The device for dynamic formation of emulsion in the flow of its components has minimal overall dimensions and a simple geometric shape in the form of a regular cylinder (for example, the device with a capacity of 50 gallons of emulsion per hour has a diameter of only 37 millimetres with a total length of only 150 millimetres).

The unit can be built in any scale factor.

The device has a micro variant (diameter 14 millimetres, length 60 millimetres).

The device can have at least eight different versions with the same overall dimensions by changing the design and dimensions of its parts and components (e.g. presence or absence of vortex generator).

Operational and installation advantages of the device.

The device for dynamic emulsion formation in the flow of its components can work in any control and monitoring scheme, including the use of electromagnetic resonance spectroscopy for control and measurement.

The device for dynamic emulsion formation in the flow of its components has no moving parts, which greatly facilitates control and monitoring in fully automatic mode. The device for dynamic emulsion formation in the flow of its components can be easily integrated into the fuel system of an internal combustion engine without requiring any modernisation of the fuel system. engine system, including rotary engines realising the thermodynamic Otto cycle. Unique properties and characteristics of the technology:

- the emulsion is formed in the developed turbulent flow of components due to the calculated combination of hydrodynamic effects, without mechanical action and without the use of any chemical activating or stabilising agents .

- time of emulsion formation does not exceed 1 second;

- The emulsion is formed in the developed turbulent flow of the base component of the emulsion, which is usually divided into two parts (e.g. 60% and 40% of the total weight of the liquid in the flows);

- The second part of the flow of the main base emulsion component is introduced into the device by means of an integrative inlet comprising at least three radial channels;

- during emulsion formation, the flow homogenisation by turbulence level occurs;

- Coaxial conical ring channels are used to homogenise the flow of the base component of the emulsion, the thickness of the flow in which does not exceed 100 microns;

- A first portion of the flow of the main base emulsion component is introduced into the outer (encompassing) conical annular channel of the device;

- the second part of the flow of the main base emulsion component is introduced into the inner (encompassed) conical annular channel of the device;

- As a rule, the flow thickness in the encompassing (outer) channel is at least two or more times greater than in the encompassing (inner) channel;

- at the outlet of each of the coaxial conical annular channels, a type of Bernoulli effect is formed in combination with cavitation and the corresponding hydrodynamic effect and with local saturation of the discontinuities resulting from the cavitation process in the flows with static electricity;

- water, as a component of emulsion, is injected into the place, where under the influence of Bernoulli effects a zone of reduced pressure is formed and flows from external and internal channels intersect;

- the design of the device allows up to eight different components to be introduced into the emulsion to be formed at the same time;

- homogenisation stage according to the turbulence level, if necessary, can be completed by the formation of a vortex tube followed by a sharp decrease in the cross-sectional area of the channel through which the emulsion is discharged from the device to the second homogenisation stage, which is carried out under high pressure (over 2000 bar);

- As a rule, the pressure in the flows of the main base component of the emulsion is the same and should be at least 3 bar;

- Alternatively, the emulsion in the device may be subjected to pressurised saturation with a gas, including an oxidising gas, such as air or oxygen;

- As a rule, the emulsion formation process is controlled by changing the pressure in the flows of emulsion components;

- the device for dynamic emulsion formation contains no moving parts and consists only of 7 original parts;

- The device may be adapted to dynamically form a micro-emulsion without the need to subject the emulsion to high pressure;

- The device can be adapted to dynamically form a nanoscale emulsion by subjecting the micro-emulsion to a short-term high pressure (more than 2000 bar);

-the device for dynamic emulsion preparation can have the following designs;

- device for stationary preparation of micro-emulsion;

- device for stationary preparation of nanoscale - emulsion

- a device for preparing emulsion in a vehicle with direct direct injection into the combustion chamber;

- A device for preparing a micro and nanoscale emulsion having an open type nozzle at the outlet;

- A device for preparing a micro and nanoscale emulsion having a closed-type nozzle at the outlet;

- the emulsion preparation device in the vehicle may be operated with an exhaust gas recirculation system.

Linear velocity of liquid emulsion components in the device directly depends on viscosity of liquid emulsion components and on pressure (at equal values of geometrical dimensions of device channels) When moving through the channels of the device liquid emulsion components gradually increase their linear velocity of movement; for each value of viscosity and pressure there are limit values of flow thickness in all channels of the device at constant dimensions of the channels to raise the linear velocity must increase the pressure; for each value of viscosity of liquid emulsion components there is a minimum size of the thickness of the flow at which increasing the pressure does not increase the speed of movement.

The device for dynamic activation of fuel mixtures has a wide range of technological possibilities.

The device, with the same dimensions of the actuating elements, can have, for different variants of the final product, different capacities. For the variant of using the device for the production of active hydrodynamic media compressible (gasified with compressed gas) emulsions, the main material for the formation of the emulsion, - for example, diesel fuel, is introduced into the device only in one inlet and on this basis, the capacity of the liquid component of the emulsion in this case will be minimal, for example, 10 gallons per hour.For variants of the device for the production of incompressible classical emulsions, such as water-in-oil, the basic material for the formation of the emulsion, for example, diesel fuel, is introduced into the device by means of two inlets, - one central one located along the longitudinal axis of the device, through which 60% of diesel fuel is introduced, intended for filling the outer conical dispersal channel of the device with a distance between the formations of the conical surfaces of the channel in, for example, - 100 microns and the second - additional.

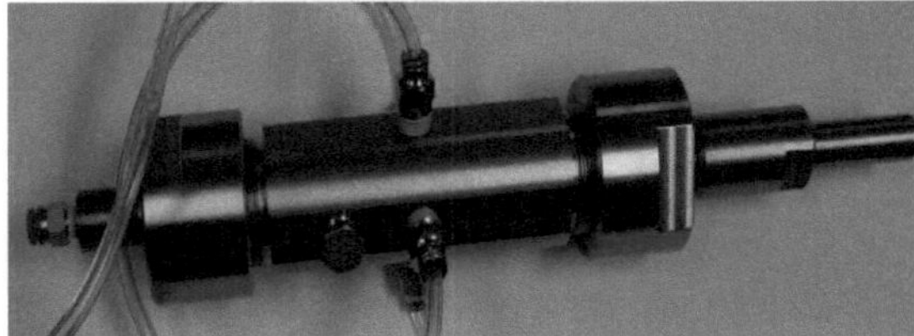

Figure 1-01, - homogenisation device, design variant.

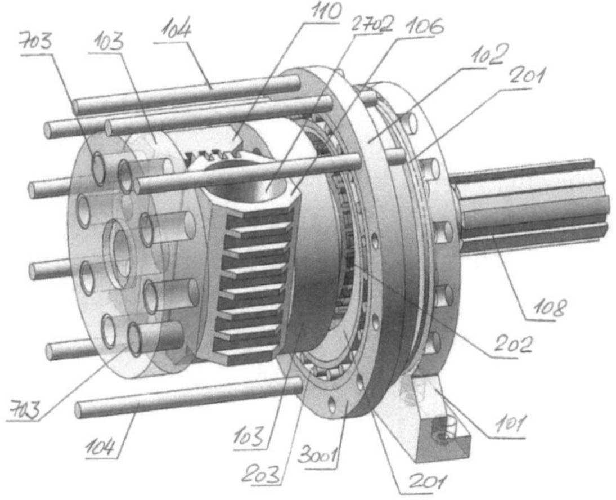

Figure 1 , - the figure shows a three-dimensional model of a rotary engine, which realises the thermodynamic Otto cycle, while the piston group of the engine and the crank mechanism do not differ from the standard ones.

The numbers in the figure denote:

101 - motor support column with eccentrically positioned support bearing;

102 - rotor flange eccentric to the axis of rotation;

103 - connecting flanges of the rotor group;

104 - axes of eccentric rotor directly connected kinematically with crank - connecting rod mechanisms;

106 - Cylinder;

108 - output shaft;

110 is a cylinder housing adjacent to the housing 106;

201 - eccentric groove;

202 - bearing;

203 - bearing;

703 - connecting and locking axles; 2702 - cylinder cavity; 3001 - bearing outer ring 203.

POTENTIAL APPLICATION OF DYNAMIC HOMOGENISATION PROCESS IN AIRCRAFT PROPULSION SYSTEMS

In connection with the recent reports on the experimental application of biological fuels or fuel mixtures for aircraft engines and knowing that fuel mixtures containing biological fuel components have the property to form clots, dynamic homogenisation of such fuels before injection into the combustion chamber can significantly increase the reliability of such engines and can open the way to the application of fuel compositions in aircraft engines;

APPLICATION OF DYNAMIC HOMOGENISATION PROCESS IN FUEL SUPPLY TO BURNERS OF BOILERS, TURBINES AND OTHER THERMODYNAMIC DEVICES

Since in the above thermodynamic systems heavier diesel fuel and different types of fuel oil are used as fuels, the formation of clots from heavier fractions with high viscosity is more intensive in such fuels. If a dynamic homogenisation device is introduced in the fuel supply and injection system of the combustion chamber, clots under certain circumstances formed in the fuel tanks and the consisting of the main hydrocarbon fraction of the fuel mixture, in the device is dynamically mixed with other hydrocarbon fractions with transformation of the mixture clots into micro or nanosized particles.Combustion of homogenised fuel occurs, as a rule, in a stable thermodynamic mode, without detonation and with a reduced content of soot and nitrogen oxides in the exhaust gases.

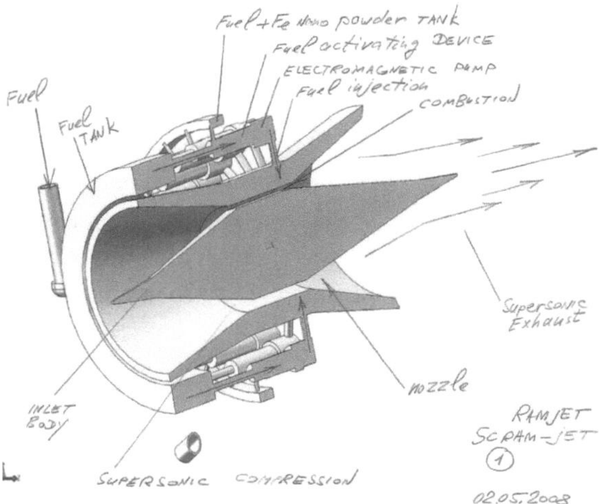

Figure 01: - a three-dimensional model of the installation of a dynamic homogenisation system on a new aircraft internal combustion engine being manufactured.

Also a three-dimensional model of the Installation of a dynamic homogenisation system on an aircraft retro internal combustion engine.

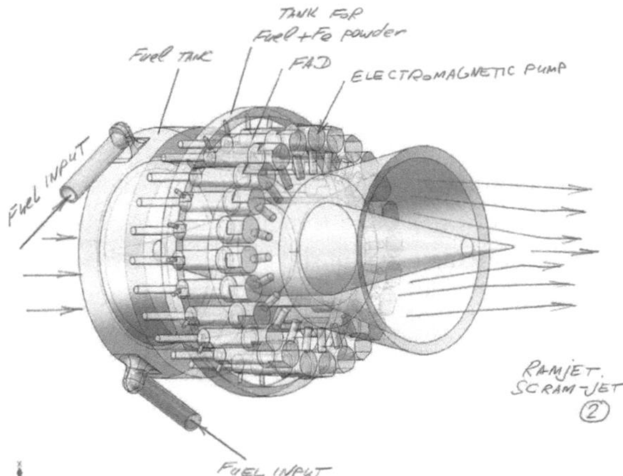

Figure 02: -also the installation of a dynamic homogenisation system on a new aircraft internal combustion engine being manufactured.

The installation of a dynamic homogenisation system on a newly manufactured aircraft internal combustion engine does not require any new elements in the engine design.

A system is introduced into the fuel line after the fuel pump, the outlet of which is connected to the inlet to the high-pressure pump or, in the absence thereof, to the inlet to the structural element following the fuel pump. Also Installation of a dynamic homogenisation system on an aircraft retro internal combustion engine.

The installation of the dynamic homogenisation system on an aircraft internal combustion engine under repair or modification does not require any new elements in the engine structure.

A system is introduced into the fuel line after the fuel pump, the outlet of which is connected to the inlet to the high-pressure pump or, in the absence thereof, to the inlet to the structural element following the fuel pump.

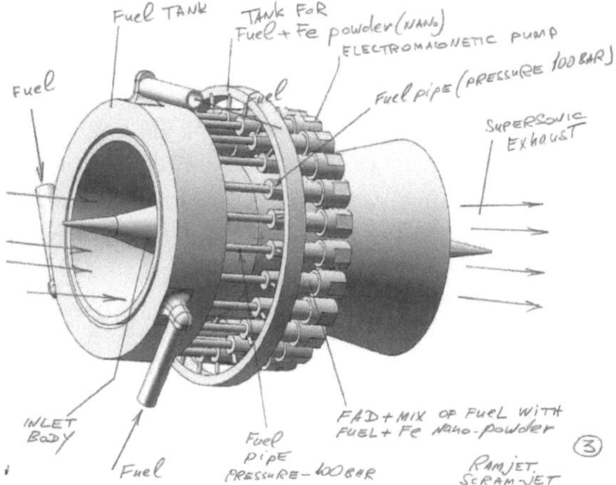

Figure 03: - also a three-dimensional model of the installation of a dynamic homogenisation system on a new aircraft internal combustion engine being manufactured.

Also a three-dimensional model of the Installation of a dynamic homogenisation system on an aircraft retro internal combustion engine.

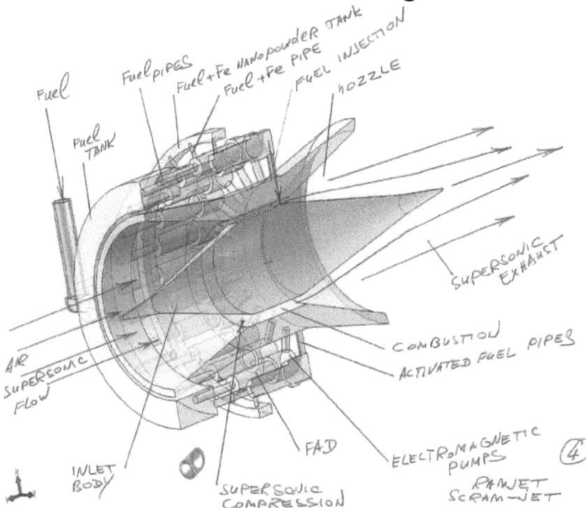

Figure 04: - also a three-dimensional model of the installation of a dynamic homogenisation system on a new aircraft internal combustion engine being manufactured.

Also a three-dimensional model of the Installation of a dynamic homogenisation system on an aircraft retro internal combustion engine.

The installation of a dynamic homogenisation system on an aircraft internal combustion engine under repair or modification does not require any new elements in the engine structure. A system is introduced into the fuel line after the fuel pump, the outlet of which is connected to the inlet to the high-pressure pump or, in the absence thereof, to the inlet to the structural element following the fuel pump.

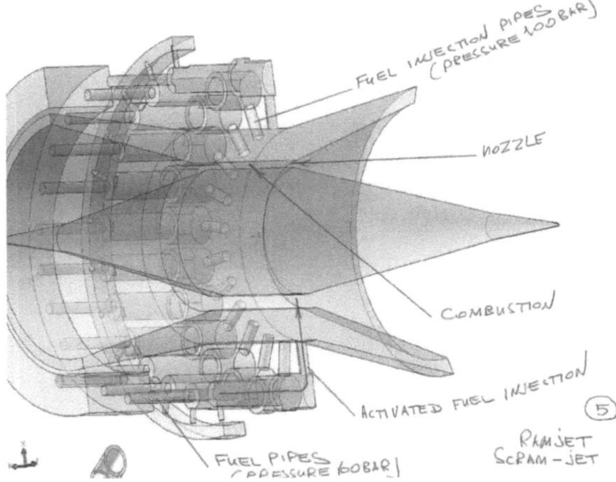

Figure 05: - also a three-dimensional model of the installation of the dynamic homogenisation system on a new aircraft internal combustion engine being manufactured.

Also a three-dimensional model of the Installation of a dynamic homogenisation system on an aircraft retro internal combustion engine. The installation of a dynamic homogenisation system on an aircraft internal combustion engine under repair or modification does not require any new elements in the engine structure.

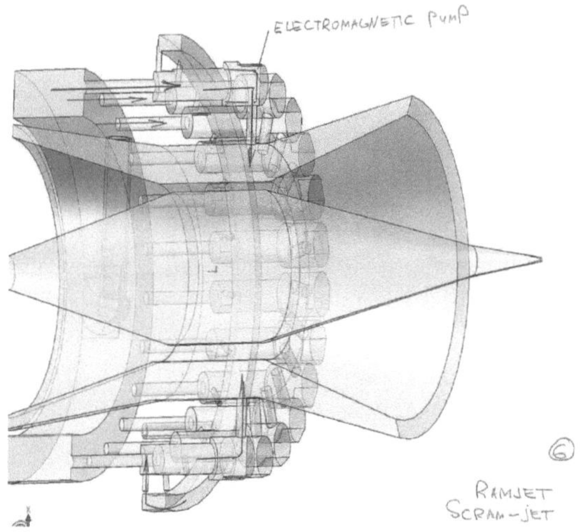

Figure 06: - also a three-dimensional model of the installation of a dynamic homogenisation system on a new aircraft internal combustion engine being manufactured.

Also a three-dimensional model of the Installation of a dynamic homogenisation system on an aircraft retro internal combustion engine. The installation of a dynamic homogenisation system on an aircraft internal combustion engine under repair or modification does not require any new elements in the engine structure.

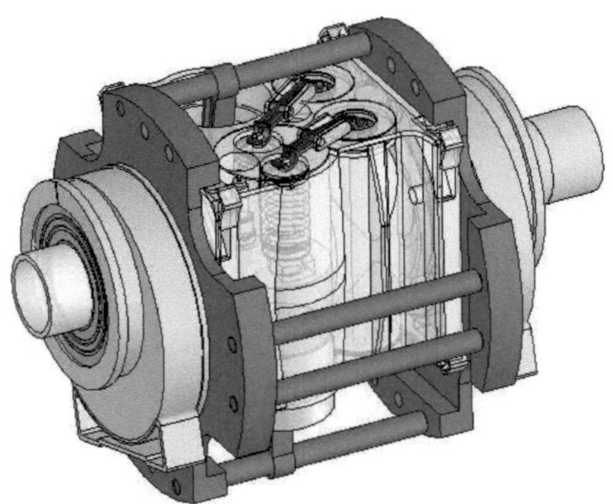

Figure 07: - the figure shows a three-dimensional model of a rotary engine that implements the Otto thermodynamic cycle, while both the piston group of the engine and the crank mechanism do not differ from the standard ones.

The proposed design variant and layout concept allow to realise the Otto thermodynamic cycle on the rotary engine and to exclude the disadvantages of rotary engines, which prevent their distribution in the models of small-sized vehicles.

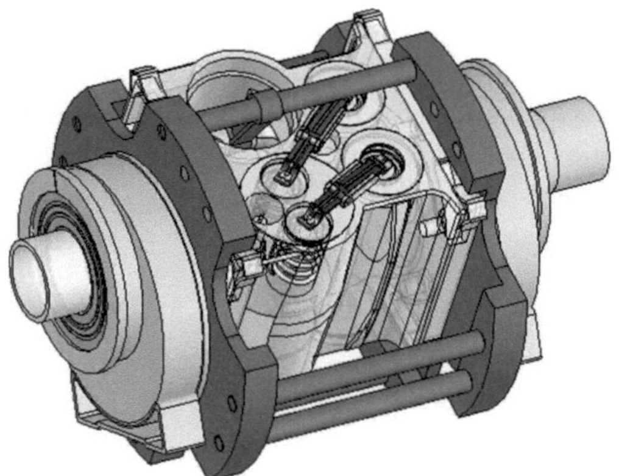

Figure 08: - the figure also shows a three-dimensional model of a rotary engine that realises the Otto thermodynamic cycle, while both the piston group of the engine and the crank mechanism do not differ from the standard ones.

The proposed design variant and layout concept allow to realise the Otto thermodynamic cycle on the rotary engine and to exclude the disadvantages of rotary engines, which prevent their distribution in the models of small-sized vehicles.

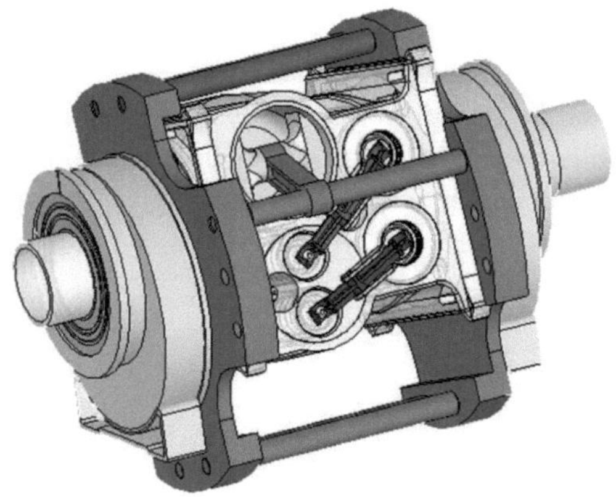

Figure 09: - the figure also shows a three-dimensional model of a rotary engine that realises the Otto thermodynamic cycle, while both the piston group of the engine and the crank mechanism do not differ from the standard ones.

As the statistics of innovation projects shows, the algorithmic component, including the logistics of the entire innovation process, starting from the formulation and synthesis of an innovative idea and ending with the process of integration into specific production and commercial structures, has an increasing impact on the commercial value and efficiency of these projects. The setting of the objective, the choice of evaluation criteria and the nature of the pathways to achieve the required results often determine the success or failure of the implementation process. As the author has experience and, technological and commercial developments in the most demanded today technologies of fuel mixture modification, he proposes as a For example, consider the algorithmisation of this group of innovative projects, which are the final stage after the multifunctional stages of transportation and delivery, usually including systemic remote monitoring, including those combined with aerial photography or using unmanned aerial vehicles. Currently, according to the information that can be obtained from public sources, there is a tendency to modify and modernise internal combustion engines in areas related to the improvement of automatic control systems for the process of fuel supply and combustion; In studies related to the use of alternative fuels such as ethanol or methanol, the problem associated with blending ethanol, methanol and petrol or ethanol, methanol and diesel clearly emerges. More than a century old problem related to the adopted mechanical system of converting linear motion into rotary motion is not considered or is considered in localised aspects not essential to the solution of the problem as a whole, and it is the mechanical problems that eat up 50% of the efficiency of any internal combustion engine.

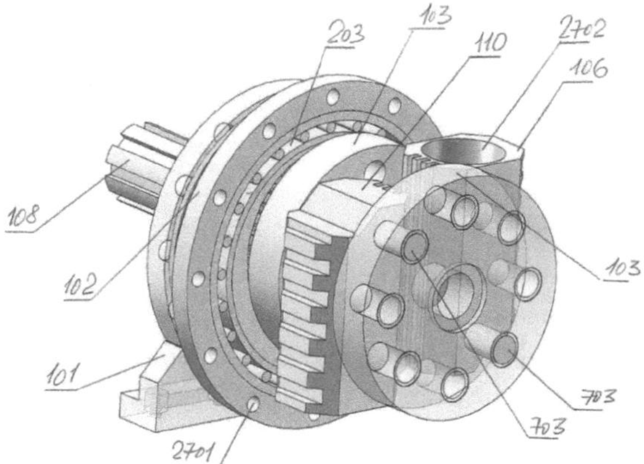

Figure 010: - the figure also shows a three-dimensional model of a fragment of a rotary engine that realises the Otto thermodynamic cycle, while both the piston group of the engine and the crank mechanism do not differ from the standard ones.

The proposed design variant and layout concept allow to realise the Otto thermodynamic cycle on the rotary engine and to exclude the disadvantages of rotary engines, which prevent their distribution in the models of small-sized vehicles.

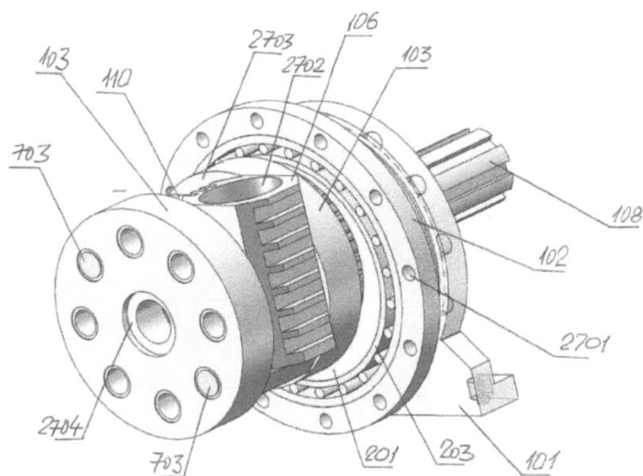

Figure 011: - the figure also shows a three-dimensional model of a fragment of a rotary engine that realises the Otto thermodynamic cycle, while both the piston group of the engine and the crank mechanism do not differ from the standard ones.

The proposed design variant and layout concept allow to realise the Otto thermodynamic cycle on the rotary engine and to exclude the disadvantages of rotary engines, which prevent their distribution in the models of small-sized vehicles.

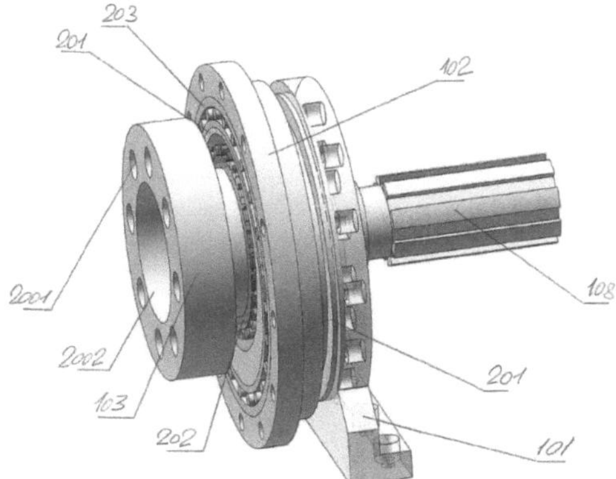

Figure 012: - the figure also shows a three-dimensional model of a fragment of a rotary engine that realises the Otto thermodynamic cycle, while both the piston group of the engine and the crank mechanism do not differ from the standard ones.

The proposed design variant and layout concept allow to realise the Otto thermodynamic cycle on the rotary engine and to exclude the disadvantages of rotary engines, which prevent their distribution in the models of small-sized vehicles.

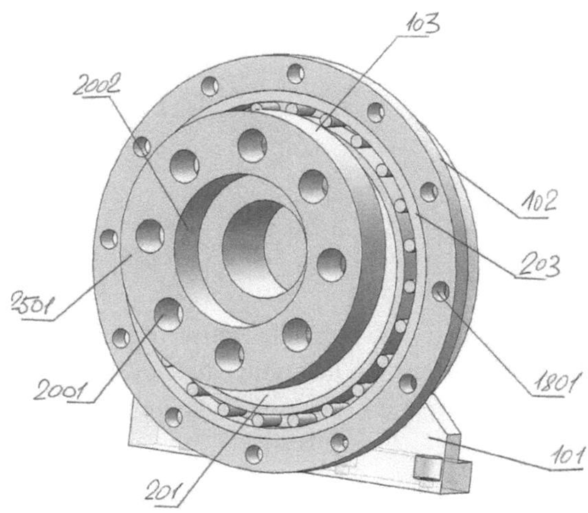

Figure 013: - the figure also shows a three-dimensional model of a fragment of a rotary engine that implements the Otto thermodynamic cycle, while both the piston group of the engine and the crank mechanism do not differ from the standard ones.

The proposed design variant and layout concept allow to realise the Otto thermodynamic cycle on the rotary engine and to exclude the disadvantages of rotary engines, which prevent their distribution in the models of small-sized vehicles.

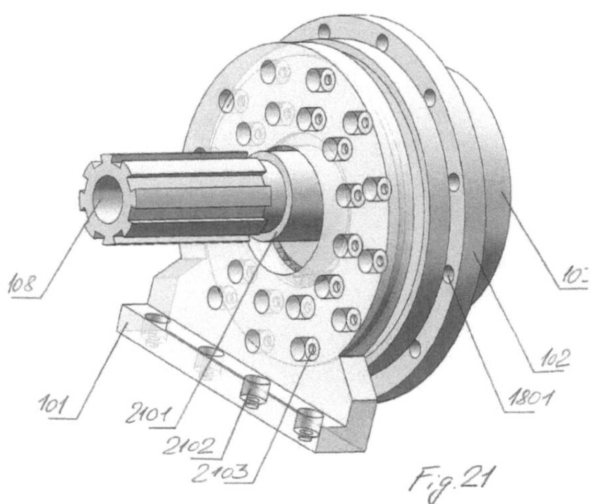

Figure 014: - the figure also shows a three-dimensional model of a fragment of a rotary engine that realises the thermodynamic Otto cycle, while the piston group of the engine and the crank mechanism do not differ from the standard ones;

The proposed design variant and layout concept allow to realise the Otto thermodynamic cycle on the rotary engine and to exclude the disadvantages of rotary engines, which prevent their distribution in the models of small-sized vehicles;

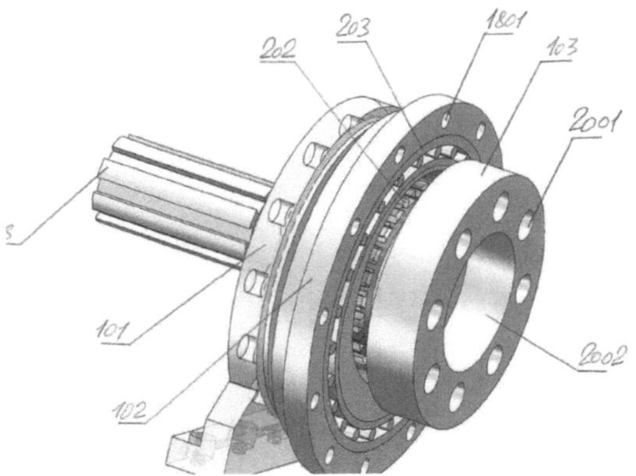

Figure 015: - the figure also shows a three-dimensional model of a fragment of a rotary engine that realises the Otto thermodynamic cycle, while both the piston group of the engine and the crank mechanism do not differ from the standard ones.

The proposed design variant and layout concept allow to realise the Otto thermodynamic cycle on the rotary engine and to exclude the disadvantages of rotary engines, which prevent their distribution in the models of small-sized vehicles.

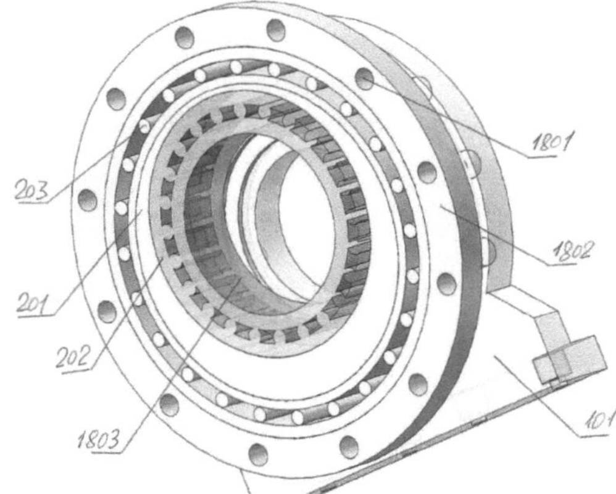

Figure 016: - the figure also shows a three-dimensional model of a fragment of a rotary engine that realises the Otto thermodynamic cycle, while both the piston group of the engine and the crank mechanism do not differ from the standard ones.

The proposed design variant and layout concept allow to realise the Otto thermodynamic cycle on the rotary engine and to exclude the disadvantages of rotary engines, which prevent their distribution in the models of small-sized vehicles.

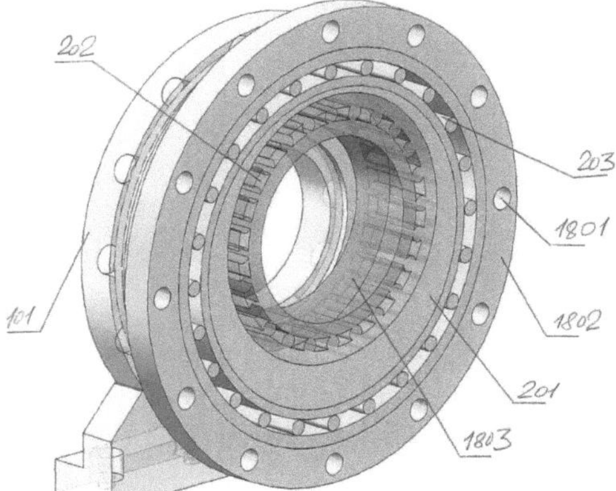

Figure 017: - the figure also shows a three-dimensional model of a fragment of a rotary engine that realises the Otto thermodynamic cycle, while both the piston group of the engine and the crank mechanism do not differ from the standard ones.

The proposed design variant and layout concept allow to realise the Otto thermodynamic cycle on the rotary engine and to exclude the disadvantages of rotary engines, which prevent their distribution in the models of small-sized vehicles.

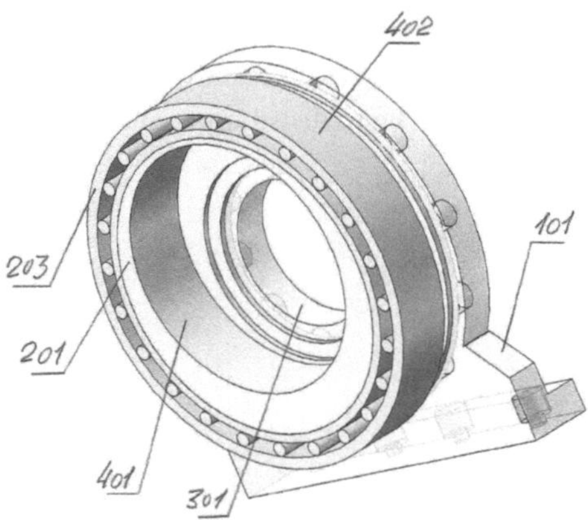

Figure 018: - the figure also shows a three-dimensional model of a fragment of a rotary engine that realises the Otto thermodynamic cycle, while both the piston group of the engine and the crank mechanism do not differ from the standard ones.

The proposed design variant and layout concept allow to realise the Otto thermodynamic cycle on the rotary engine and to exclude the disadvantages of rotary engines, which prevent their distribution in the models of small-sized vehicles.

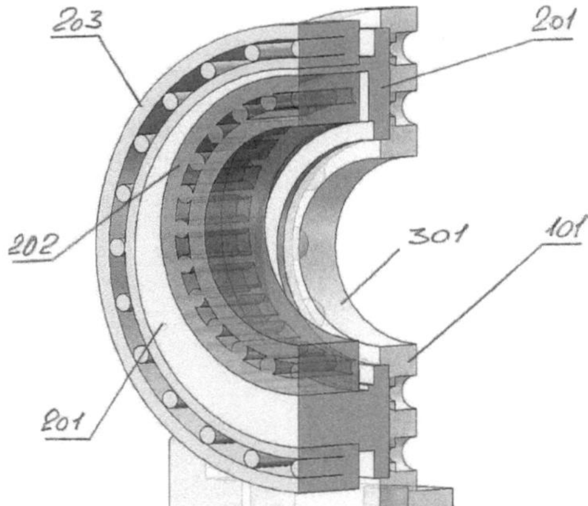

Figure 019: - the figure also shows a three-dimensional model of a fragment of a rotary engine that realises the Otto thermodynamic cycle, while both the piston group of the engine and the crank mechanism do not differ from the standard ones.

The proposed design variant and layout concept allow to realise the Otto thermodynamic cycle on the rotary engine and to exclude the disadvantages of rotary engines, which prevent their distribution in the models of small-sized vehicles.

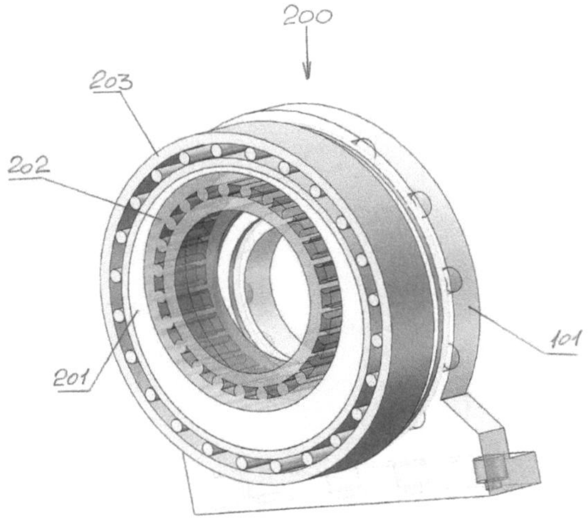

Figure 020: - the figure also shows a three-dimensional model of a fragment of a rotary engine that realises the Otto thermodynamic cycle, while both the piston group of the engine and the crank mechanism do not differ from the standard ones.

The proposed design variant and layout concept allow to realise the Otto thermodynamic cycle on the rotary engine and to exclude the disadvantages of rotary engines, which prevent their distribution in the models of small-sized vehicles.

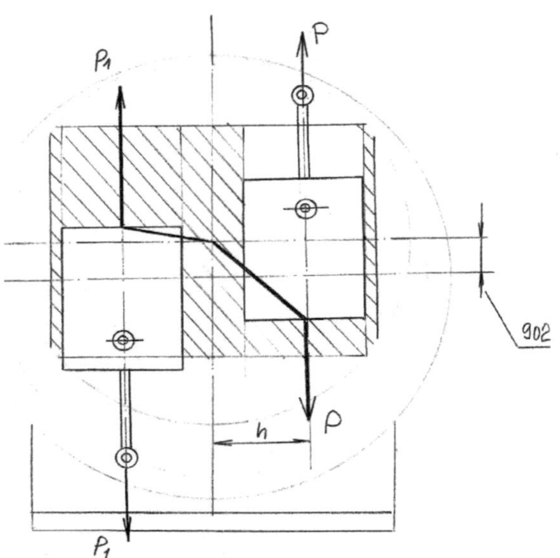

Figure 021: - the figure also shows the circuit diagram and three-dimensional model of a fragment of a rotary engine, which implements the thermodynamic Otto cycle, while the piston group of the engine and crank mechanism do not differ from the standard.

The proposed design variant and layout concept allow to realise the Otto thermodynamic cycle on the rotary engine and to exclude the disadvantages of rotary engines, which prevent their distribution in the models of small-sized vehicles.

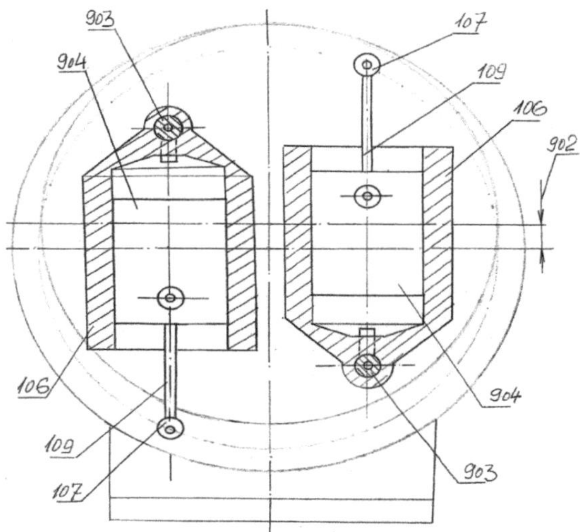

Figure 022: - the figure also shows the circuit diagram and three-dimensional model of a fragment of a rotary engine, which realises the thermodynamic Otto cycle, while both the piston group of the engine and the crank mechanism do not differ from the standard ones.

The proposed design variant and layout concept allow to realise the Otto thermodynamic cycle on the rotary engine and to exclude the disadvantages of rotary engines, which prevent their distribution in the models of small-sized vehicles.

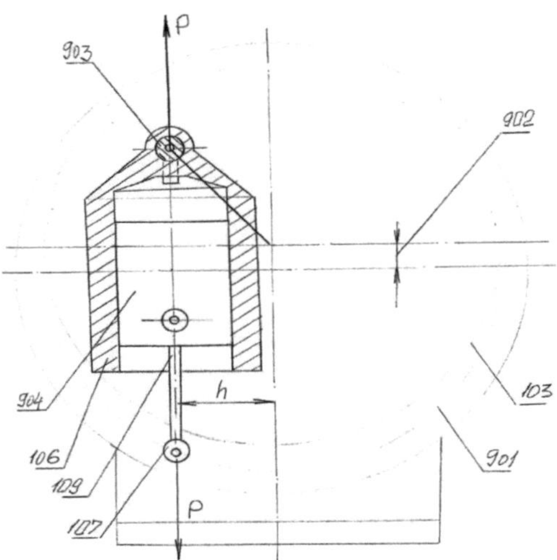

Figure 023: - the figure also shows the circuit diagram and three-dimensional model of a fragment of a rotary engine, which realises the thermodynamic Otto cycle, while both the piston group of the engine and the crank mechanism do not differ from the standard ones.

The proposed design variant and layout concept allow to realise the Otto thermodynamic cycle on the rotary engine and to exclude the disadvantages of rotary engines, which prevent their distribution in the models of small-sized vehicles. The author's developments include filed applications for inventions that address the set of problems described above. The problem of mixing ethanol and petrol at petrol stations is completely solved by the invention, which has a prototype made for butter churning, but the same prototype and the basic principles of its operation can be used to present the idea and solve the problem of mixing and turning the mixture into a stable emulsion, when mixing fuel components, such as diesel fuel and various types of technical alcohol, including glycerin; ethanol and petrol and many others, including those of inorganic origin. In addition to all the systems and methods of the various automatic control systems known from publications, the author proposes a system for non-contact monitoring of the actual state of the fuel mixture, including the level of its saturation with air and the level and nature of foaming of the fuel mixture before it is injected or fed to a high-pressure fuel pump, as is the case in diesel engines; the application for the invention was filed last year. At the same time, the author notes that one of the most important conditions for the successful commercial application of these innovations are the conditions that allow the use of the most sophisticated remote control and monitoring methods for online control and monitoring, combined with aerial photography processes or control or monitoring by unmanned aerial vehicles.The author co-operates with a creative group that has proposed a solution to mechanical problems in all types of internal combustion engines by modifying the design of the mechanism for converting linear movement into rotary, without any changes to the fuel system of the internal combustion engine; the application for the invention was filed last April.

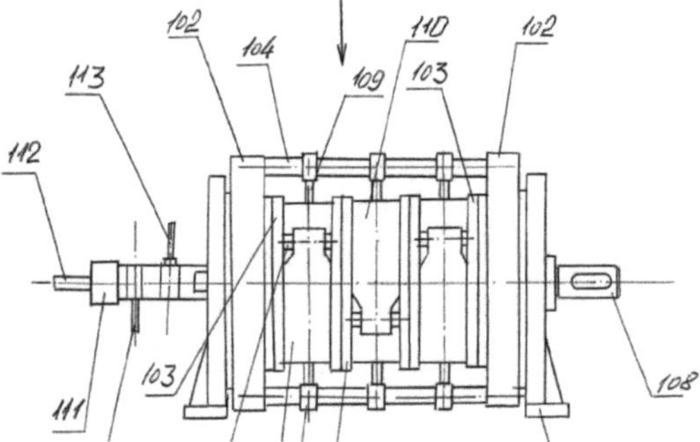

Figure, rotary internal combustion engine with integrated fuel or fuel mixture homogenisation system.

The numbers in the figure denote:

102- flanges of the rotor part of the motor kinematics

103- sections for mounting engine cylinders

104- axles for outputtingcrank -connecting rod mechanisms of engine cylinders

108- motor output shaft

109- connecting rods of crank - connecting rod mechanisms of the engine

110- engine cylinders

111- device for on-line homogenisation of fuel mixtures or mono-fuels

112- introduction of fuel or fuel mixture into the device for on-line homogenisation of fuel mixtures or mono - fuel

113 - radial inlets of air or fuel mixture components or fuel emulsion components into the device for on-line homogenisation of fuel mixtures or mono - fuels

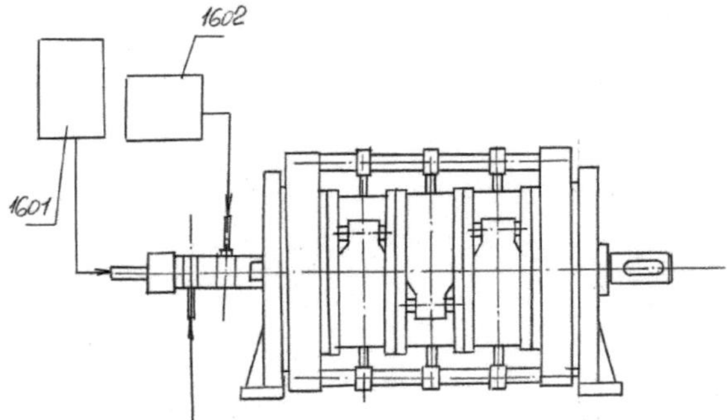

Figure 3 - rotary internal combustion engine with integrated fuel or fuel mixture homogenisation system, showing the functional interaction with the liquid fuel tank and with the fuel component tank or with the compressed air compressor.

The numbers in the figure denote:

1601 - multifunctional tank, for the option with on-line homogenisation of mono-fuels, or as an option with on-line homogenisation of fuel blends (with re-mixing system built into the tank), or as an option with on-line homogenisation of fuel emulsions (with re-mixing system built into the tank); The system shall also include a low pressure pump and all associated control and monitoring and measuring equipment;

1602 - A multifunctional tank, for the embodiment of giving mixing and homogenisation functions to the device, wherein the ethanol blend contains ethanol, wherein the methanol blend contains methanol, wherein the fuel emulsions contain water;

The system shall also include a low pressure pump and all associated control and monitoring - measuring equipment; In the case of the gasified fuel option, this is the compressor with all the associated control and monitoring and measuring equipment.

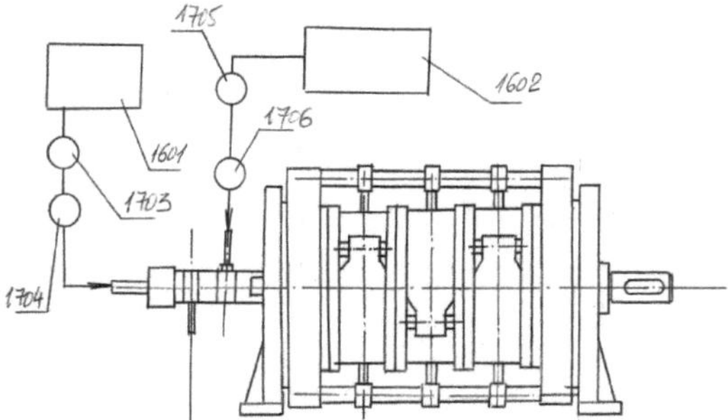

Figure 4 - rotary internal combustion engine with integrated fuel or fuel mixture homogenisation system with indication of functional interaction with liquid fuel tank and fuel component tank or with compressed air compressor and all associated control and monitoring - measuring equipment.

The numbers in the figure denote:

1601 - multifunctional tank, for the option with on-line homogenisation of mono-fuels, or as an option with on-line homogenisation of fuel blends (with re-mixing system built into the tank), or as an option with on-line homogenisation of fuel emulsions (with re-mixing system built into the tank); The system shall also include a low pressure pump and all associated control and monitoring and measuring equipment;

1602 - A multifunctional tank, for the embodiment of giving mixing and homogenisation functions to the device, wherein the ethanol blend contains ethanol, wherein the methanol blend contains methanol, wherein the fuel emulsions contain water;

1703 - pump with a system of valves, check valves, pressure gauges, pressure regulators, pressure stabilisers, pulsation stabilisers and, as a version - electromagnetic - resonance flow sensor - composition of fuel mixture components;

1704 - flow meter with a system of check valve valves, pressure gauges, pressure regulators, pressure stabilisers, pulsation stabilisers and as a version - electromagnetic - resonant flow sensor - composition of fuel mixture components;

1705- pump with a system of valves, check valves, pressure gauges, pressure regulators, pressure stabilisers, pulsation stabilisers and, as a version - electromagnetic - resonant flow sensor - composition of fuel mixture components;

1706 - flow meter with a system of check valve valves, pressure gauges, pressure regulators, pressure stabilisers, pulsation stabilisers and as a version - electromagnetic - resonant flow sensor - composition of fuel mixture components;

All of the above technologies can be applied if a number of fundamental conditions are fulfilled.

To ensure timely and complete combustion in a short period of time, the fuel must fulfil the following requirements:
1) have good conductivity to ensure reliable operation of the high-pressure fuel pump (at optimum viscosity of 2-6 mm2/s and temperature of 20 °C); good low-temperature properties; absence of mechanical impurities and water;

Good permeability is the most important factor enabling on-line non-contact monitoring of all types of liquid transport parameters in pipelines in parallel with aerial survey operations or by remote control and monitoring methods using unmanned aerial vehicles.

2) provide the necessary atomisation, good mixing and vaporisation; for this purpose, the fuel must have an optimum viscosity and a certain fractional composition;

3) have the necessary flammability to ensure easy starting of a cold engine, smooth pressure build-up and complete smokeless combustion (these properties depend on the chemical and fractional composition of the fuel, as well as viscosity; the chemical composition of the fuel is assessed by the octane number, which characterises flammability and is the main indicator of the motor properties of the fuel);

4) not to cause increased formation of carbon deposits and other deposits on valves, rings, pistons, coating of atomiser needle with coke (propensity of fuel to carbon deposits depends on chemical and fractional composition, viscosity, content of mechanical impurities and water);
5) do not contain corrosive products (corrosive properties of fuel depend on the

presence of mineral and organic acids, sulphur compounds and water);

6) to have the highest possible calorific value;

7) to have high fuel permeability with minimum leakage through the gaps in the plunger pairs with minimum wear of the rubbing pairs.

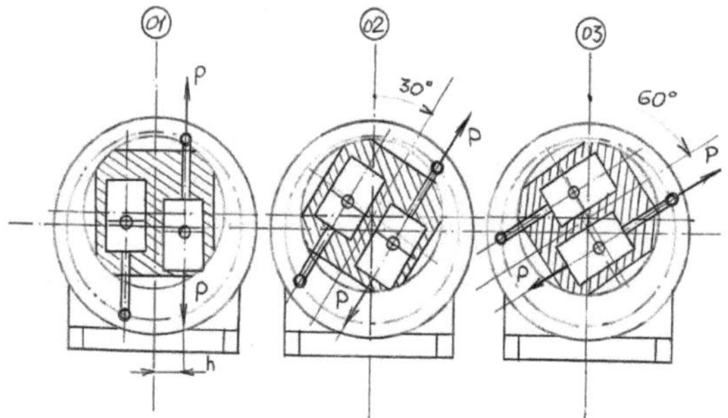

Figure 5, - the figure shows the first three positions of the rotor with cylindrical - piston group rotor kinematics every 30 degrees of rotor rotation.

As can be seen from the diagrams, the presence of eccentricity in the installation of the rotor allows to exclude the so-called dead zones and in turn allows to obtain torque even at points of coincidence of the direction of the connecting rod of the crank - connecting rod mechanism with the vertical or horizontal axis, in which in conventional engines torque is zero.

Now, to imagine that it is the engine of a pumping or compressor station on a main pipeline, it is necessary to exclude any fluctuations of the controlled liquids or mixtures in order to make it possible to conduct a realistic monitoring of the flow parameters of liquid products in the pipeline.

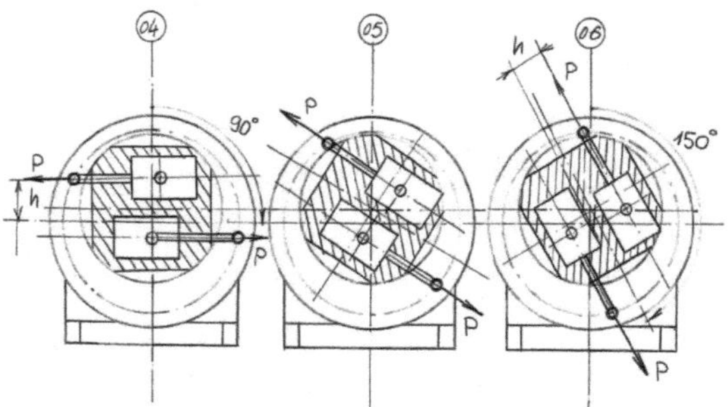

Figure 6 - the figure shows three successive positions of the rotor with cylindrical - piston group of rotor kinematics every 30 degrees of rotor rotation, from 90 degrees of rotation to 150 degrees of rotor rotation. of the rotor.

As can be seen from the diagrams, and in the subsequent stages of rotation, the presence of eccentricity in the rotor setting allows to exclude the so-called dead zones and in turn allows to obtain torque even at points of coincidence of the direction of the connecting rod of the crank - connecting rod mechanism with the vertical or horizontal axis, in which in conventional engines torque is zero.

The option of using gaseous fuel instead of liquid fuel in the rotary engine is extremely interesting. An analytical study of the development of a chemical reaction accompanied by heat release in a laminar gas flow and under intensive mixing of reaction products with initial substances has revealed the hysteresis character of the solutions of the combustion theory; by increasing, for example, the initial temperature of a combustible gas, it is possible to reach the moment when the gas will ignite; but to extinguish a gas burning at a high temperature, it is possible to use a hysteresis solution. temperature of the gas, it is necessary to reduce the initial temperature significantly below that at which ignition occurred. This example of operation of a chemical reactor of ideal mixing illustrates also a fundamental property of the combustion process - the possibility of existence of several stationary modes of combustion at the same given external parameters. This ambiguity essentially distinguishes stationary processes with continuous supply of initial substances and withdrawal of reaction products from a static thermodynamically equilibrium situation.

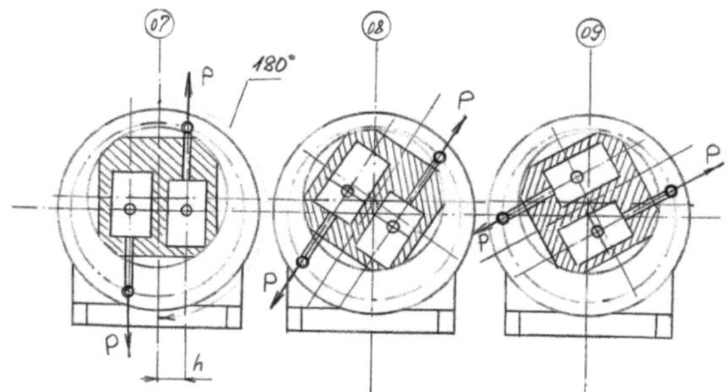

Figure 7 - the figure shows the next three following rotor positions after 150 degrees of rotor position with cylindrical - piston group rotor kinematics every 30 degrees of rotor rotation, from 180 degrees o f rotor rotation to 270 degrees of rotor rotation

As can be seen from the diagrams, and in the subsequent stages of rotation, the presence of eccentricity in the rotor setting allows to exclude the so-called dead zones and in turn allows to obtain torque even at points of coincidence of the direction of the connecting rod of the crank - connecting rod mechanism with the vertical or horizontal axis, in which in conventional engines torque is zero.

The stability of the hydrodynamic or aerodynamic parameters of the proposed rotary engine and its local infrastructural features opens up new opportunities for efficiency improvements, primarily in avoiding geometry-uncertain velocity pulsations.As determined, velocity pulsations directed across the direction of combustion propagation cause asymmetry of flame front curvature, contribute to its turns and formation of complex structure. The main task of the theory of turbulent combustion is the study of a stationary on average turbulent flame: in a given turbulent flow field it is required to find the time-averaged structure of the combustion zone and to determine its statistical characteristics: the average combustion velocity, the average surface of the flame front, the width of the area occupied on average by the curved flame front, and other characteristics. An important role in this theory should be assigned to the leading points of the curved front - laminar flame, carried by turbulent pulsations towards the combustible gas, as they ensure the existence of the flame surface following them and determine the average combustion velocity.The velocity pulsations of a turbulent gas flow under developed turbulence often exceed the normal flame propagation velocity. The rate of leading point removal into the flammable gas is determined mainly by the pulsation characteristics of the flow. Thus, the prevention of velocity pulsations becomes even more important.

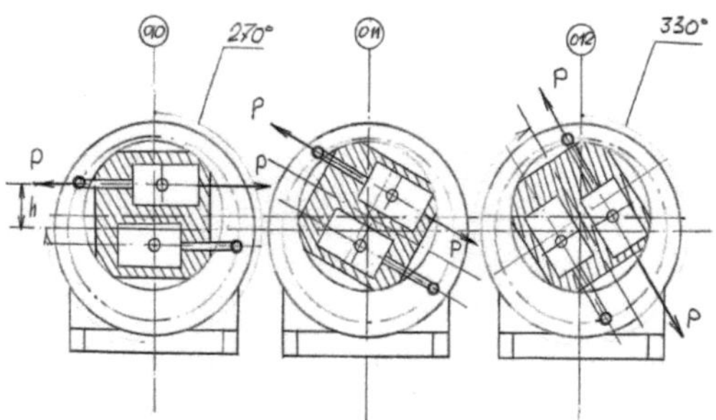

Figure 8 - the figure shows the next three rotor positions after 150 degrees of rotor position with cylindrical - piston group rotor kinematics every 30 degrees of rotor rotation, from 270 degrees o f rotor rotation to 330 degrees of rotor rotation.

As can be seen from the diagrams, and in the subsequent stages of rotation, the presence of eccentricity in the rotor setting allows to exclude the so-called dead zones and in turn allows to obtain torque even at points of coincidence of the direction of the connecting rod of the crank - connecting rod mechanism with the vertical or horizontal axis, in which in conventional engines torque is zero.

Previously, it was believed that fuel for high-speed engines should have a viscosity of at least 5 mm2/s at 20 °C. Studies have shown that fuel with viscosity up to 2 mm2/s at 20 °C provides lubrication of fuel supply equipment. The minimum viscosity depends on the injection pressure and other design solutions. With increasing injection pressure fuel viscosity can increase by 6-10 times. (use of the technology of complex activation of fuel mixtures in the form of complex technology of emulsion formation by the water-in-oil method, in which water is mixed with diesel fuel, forming a stable emulsion, which is then mixed with air bubbles in which internal pressure in the bubbles is from 10 to 20 bar, allows almost halving the required pressure, which reduces the viscosity of the fuel mixture by 3 - 5 times; in addition, an admixture of 10% water and 8 - 12% air in the same proportion reduces the initial viscosity of the fuel mixture). On the basis of researches and operational tests the following values of fuel viscosity at 20°C for high-speed diesel engines have been established: in summer - 3.0-8.0 mm2/s, in winter - 2.2-6.0 mm2/s, for the severe climate of the Arctic - 1.5-4 mm2/s.The conductivity of the fuel also depends on its low-temperature properties, which affect the mobility of the fuel at low temperatures. Low-temperature properties are determined by the turbidity, crystallisation and solidification temperatures. The cloud point is the temperature at which the phase homogeneity of the fuel is lost. It becomes cloudy due to the release of tiny water droplets, solid hydrocarbons or microscopic ice crystals. The temperature at which the first crystals visible to the naked eye appear is called the crystallisation temperature. The temperature of complete loss of mobility is called the solidification temperature. (For the proposed technology, the solidification temperature of the mixture drops in proportion to the percentage of water and air in the mixture, and, in addition, the high level of turbulence of the mixture also contributes to lowering the solidification temperature).

Mechanical impurities in fuel are not stipulated by the standard (not allowed). Fuel is contaminated if the rules of transport, storage and refuelling are not observed. Alumina quartzite is the most harmful as it has high hardness. Precision pairs of fuel pumps have clearances of 1.5-3.0 microns, so even a small percentage of mechanical impurities leads to significant abrasive wear. (when using the proposed technology, the ingress of abrasives into the fuel pump is excluded, thanks to a special system of capillary channels in the device for complex activation of diesel fuel; In these devices coaxial channels of hydrodynamic interface have sizes from 25 microns to 100 microns). Dissolved water is always present in the fuel to a greater or lesser extent. Its concentration depends on the ambient temperature. Water solubility in fuel is 10~5 kg/kg. Especially unpleasant is the presence of emulsion water at low temperatures. In this case, ice crystals clog the cleaning system and disrupt engine operation. Water has a negative effect on fuel equipment, especially on high-pressure pump and injectors. It promotes corrosion of surfaces of precision couples and coating of injector atomisers with coke.

(in the proposed technology the formation of ice crystals is impossible, and, in addition, the mixture can have a special section of on-line heating, which eliminates the occurrence of ice crystals, reduces or completely eliminates the occurrence of corrosion and the formation of coke in the nozzle atomisers). The required atomisation, mixing and vaporisation of the fuel largely predetermine the working process as a whole, its efficiency and economy. A portion of fuel measured by a high-pressure pump is injected into the combustion chamber into dense, strongly swirled heated air. At some distance from the nozzle openings of the nozzle, the jet breaks up into droplets, forming a plume of atomised fuel. The total number of droplets reaches several million and their size varies from 50 to 150 microns. The distribution of droplets by number and size is very uneven. The quality of fuel atomisation is characterised by the number and size of droplets, length, width and angle of the spray cone.

Fuel droplets introduced into hot air do not ignite instantly. At ignition, the vaporisation process is more intense due to the high temperature of the combustion process. At the same time with acceleration of vaporisation at this moment there is some slowing down, because there are combustion products that hinder the supply of air oxygen to the vaporising fuel. This circumstance makes it necessary to carry out the working process in the diesel engine with some excess air.

(the proposed technology allows to ensure the necessary excess of air and high uniformity of air distribution in the volume of the fuel mixture, in addition, three-dimensional encapsulation of the fuel mixture, when capsules are spheres having a core of air with a shell of fuel, allow to obtain at injection separate vaporisation of fractions, which makes it possible, when air expands after injection, to break capsule shells into particles in the submicron range, which sharply improves the efficiency of combustion process in engine cylinders).The large heterogeneity of the fuel-air mixture in the combustion chambers is the reason for some advantages and disadvantages of diesel engines. An important advantage is that in diesel engines it is possible to significantly lean the working mixture. This allows the power output to be varied by the fuel supply alone. On the other hand, the inhomogeneity of the mixture is a significant disadvantage of diesel engines, as smokeless and complete combustion cannot be achieved. (the proposed technology allows to obtain high homogeneity of the mixture, which can be expressed by the size of compressed air bubbles obtained during activation - 20 micrometres). The fuel viscosity affects the mixing process, the increase of which leads to deterioration of atomisation and vaporisation of fuel. At high viscosity, large droplets increase the flame length and get on the walls, which significantly worsens the mixing process, and at low viscosity the fuel flame is shortened and the combustion chamber is not fully utilised. (the proposed technology allows to obtain particle sizes of 3 - 5 microns, which increases the completeness of combustion; the viscosity of the mixture is reduced in proportion to the percentage of impurities and the mixture is formed using the energetic advantages of the Bernoulli effect, which improves the process).The fractional composition of diesel fuels is estimated by the boiling end temperature: summer - 360 °C, winter - 340 °C, Arctic - 330 °C. The complex processes of combustion and fuel mixing in high-speed engines occur in a very short period of time, about 10 times faster than in carburettor engines, at the same speed. The intensity of combustion depends on many factors: pressure and temperature of compressed air, concentration of fuel vapours in the air, chemical composition, quality of atomisation and vaporisation of fuel.If the fuel ignition is delayed, the subsequent combustion process is very intensive. Characteristic engine knocking is heard (similar to detonation, but the causes are different).

Other things being equal, a shorter ignition delay period results in smoother pressure changes, i.e. smoother engine operation. However, excessive shortening of this period leads to a reduction in combustion completeness. The process starts immediately after the fuel is fed, most of which is fed into the combustion products. The fuel droplets vaporise rapidly before reaching the areas of the chamber where oxygen has not yet been used. The process of mixture formation is sharply deteriorated, the engine power and efficiency drops.In order to ensure a normal combustion process, it is necessary to use fuel with an optimal ignition delay period. The flammability of a fuel is assessed by its octane rating. Numerically, the octane number of diesel fuel is equal to the percentage content (by volume) of octane in the mixture with alpha-methyl naphthalene, which by the nature of combustion (self-ignition) is equal to the tested fuel. The octane number is determined by three methods: critical compression ratio, autoignition delay and flash coincidence.

Diesel fuels should have an octane rating between 45-50 in winter and 40-45 in summer. The octane rating determines the starting properties of diesel fuels that differ slightly in fractional composition.

When operating diesel engines, it is of great importance to set the optimum fuel injection advance angle. If the advance angle is large, fuel is fed into insufficiently heated air, which increases the ignition delay period and stiffness of operation. The fuel may burn to top dead centre, resulting in a loss of power as back pressure is created. With delayed injection, a significant portion of the fuel burns in the expansion line, causing a drop in power, incomplete combustion, and reduced engine efficiency.

Carbon fouling in high-speed diesel engines leads to engine overheating, coke coating of injectors, deterioration of fuel atomisation. Incomplete combustion of fuel, presence of high-molecular complex substances and mechanical impurities in fuel contribute to increased accumulation of soot. The accumulation of resinous substances is significantly affected by fuel stability. Diesel fuel quality indicators affecting soot formation and standardised by standards are as follows: coke number, tar content, ash, mechanical impurities and sulphur compounds. Sulphur contained in fuel affects not only the mass of the formed carbon black, but also its properties. Sulphur compounds, accumulating in the soot, increase its density.

Corrosive properties of diesel fuels are determined by the content of sulphur compounds, water-soluble acids and alkali, as well as water. The presence of sulphur compounds in the fuel is checked with a polished electrolytic copper plate of 10×25 mm. This plate is lowered into a porcelain bowl with fuel, which is placed in a drying cabinet at a temperature of 50 ° C and kept there for 2-3 hours. Appearance of dark brown, grey or black deposits on the plate indicates the presence of active sulphur compounds in the fuel. Corrosion of parts is mainly caused by sulphur compounds. Liquid corrosion is especially strong in cold season during starting modes. Increase of sulphur content in fuel from 0.2 to 0.5% increases wear of cylinder piston group by 25-30%, up to 1% - 2 times. To reduce sulphur corrosion, additives are added to fuel. The most common one is zinc naphthenate, which is added to fuel (0.25-0.30% of fuel mass). (the proposed technology completely eliminates the presence of sulphur compounds in the mixture and, if necessary, allows to increase the temperature of the mixture, which reduces corrosion of diesel engine parts by 80 - 85%).

The calorific value of diesel fuel is 42,705 kJ/kg.

Aspects of corrosion phenomena and corrosion processes in structural elements of diesel engines using a fuel mixture in which diesel fuel is mixed with compressed air. When diesel fuel is mixed with compressed air, without water, corrosion processes are excluded in the same way as in the previous case; (the proposed technology has two main versions; the first version is intensive mixing of diesel fuel with water and subsequent uniform homogeneous mixing of the mixture of diesel fuel and water with compressed air; after mixing with compressed air, the mixture has air bubbles with a diameter of 20 micrometres and an internal pressure of 20 bar; each bubble has a 20 micrometre thick shell consisting of an emulsion of 92% diesel fuel and 8% water; after injection into the combustion chamber, the mixture is atomised into particles of 3 to 5 micrometres in size, and the diesel fuel and oxidant represent a medium in which the organic combustible element is evenly distributed between the oxidant, which favours ignition at lower temperatures and creates an octane number equivalent of 45 to 50;

The second version of the technology is the blending of only diesel fuel with compressed air and, while the conditions of the first version of the technology are maintained, the ignition of the fuel mixture and the octane number equivalent retain similar qualities to the first version;)

It is well known that both ethanol and methanol bring water into the mixture after mixing - ethanol more, methanol less, but its presence cannot be discounted and therefore emulsion production becomes an extremely important process.

LIST OF USED LITERATURE, PATENT AND LICENCE MATERIALS

Annex 1

United States Patent Application 20140318123 Kind Code A1
Cruz; Jose LopezOctober 30, 2014

ROTARY INTERNAL ***COMBUSTION ENGINE***

Abstract

A **rotary engine rotary engine** according to the present invention comprises a main housing assembly and a rotor assembly rotatably supported within the housing. The rotor assembly has two rotors, an intake/compression rotor rotatably disposed within the intake/compression housing, and a power/exhaust rotor rotatably disposed within the power/exhaust housing. The rotors have N number of apexes and sides, wherein N is an integer greater than 2. A rotating chamber is formed between each side of the rotor and the inner wall of the respective housing. The stages of the thermodynamic cycle of the **engine** occur within these chambers. For example, if the rotors have three sides, the rotors will have a triangular-like shape with three apexes. The apexes form the outermost radial part of the rotors which engage the inner wall of the respective housing bore. Each of these chambers is split into two divided chambers by a reciprocating vane, thereby forming 2 times N divided chambers in each of the respective housing bores.

Annex 2

United States Patent Application 20150152781 Kind Code A1
Oledzki; Wieslaw JulianJune 4, 2015

Rotary two-stroke internal ***combustion engine,*** *particularly very high power very high speed* ***engine*** *fuelled by solid particulate fuel like coal dust, destined for power generation*

Abstract

Rotary positive displacement internal **combustion** two stroke **engine** with only one major moving part, i.e. **engine's** eccentric shaft, with cylinders encompassing eccentrics of the shaft, with said eccentrics being suitably sealed in said cylinders, wherein **combustion** gases exert force directly on said eccentrics of said eccentric shaft. Gas forces can be nullified by suitably phasing shaft's eccentrics, thus nullifying gas forces loading shaft's bearings. The **engine** is naturally perfectly balanced. The **engine** possesses natural self-cleaning capability, and is capable of being fueled by coal dust. Units of the type developing 2 000 MW and 3 000-3 6000 rev/min, intended to replace steam turbines in power stations, can be built.

Annex 3

United States Patent Application20180066520 Kind CodeA1
Shkolnik; Alexander ; et al. March 8, 2018

Rotary ***Engine***

Abstract

A **rotary engine** includes an intake port, an exhaust port, a rotor having an intake channel and/or an exhaust channel, and a rotor shaft coupled to the rotor. The rotor shaft has an inflow channel in communication with the intake channel and/or an outlet channel in communication with the exhaust channel. The rotary engine includes a housing having a working chamber formed between the housing and the rotor, the working chamber configured to handle, in succession, an intake phase, a compression phase, a **combustion** phase, an expansion phase, and an exhaust phase. The inflow channel cyclically communicates with the intake port and forms a passage between the intake port and the working chamber through the rotor shaft and the intake channel. The outlet channel cyclically communicates with the exhaust port and forms a passage between the exhaust port and the working chamber through the rotor shaft and the exhaust channel.

Annex 4

United States Patent Application 20160341042 Kind Code A1
Shkolnik; Alexander ; et al. November 24, 2016

Rotary ***Engine***

Abstract

A **rotary engine** includes a housing having a working cavity, a shaft, the shaft having an eccentric portion, a rotor having a first axial face, and a second axial face opposite the first axial face, the rotor disposed on the eccentric portion and within the working cavity, the rotor comprising a first cam on the first axial face, the first came having an eccentricity corresponding to the eccentricity of the eccentric portion of the shaft, and a cover integral with, or fixedly attached to, the housing, the cover comprising a plurality or rollers, each roller engaged with the cam, wherein the cam guides the rotation of the rotor as the rotor rotates within the working cavity and orbits around the shaft.

Annex 5

United States Patent Application 20180010456 Kind Code A1
GAUVREAU; Jean-Gabriel ; et al.January 11, 2018

INTERNAL ***COMBUSTION ENGINE*** *WITH ROTOR HAVING OFFSET PERIPHERAL SURFACE*

Abstract

A **rotary engine** where the rotor cavity has a peripheral inner surface having a peritrochoid configuration defined by a first eccentricity and the rotor has a peripheral outer surface having a peritrochoid inner envelope configuration defined by a second eccentricity larger than the first eccentricity. Also, a **rotary engine** wherein the rotor cavity has a peripheral inner surface having a peritrochoid configuration defined by an eccentricity, and a rotor with a peripheral outer surface between adjacent ones of the apex portions being inwardly offset from a peritrochoid inner envelope configuration defined by the eccentricity. The **engine** may have an expansion ratio with a value of at most 8. The **rotary engine** may be part of a compound **engine** system.

Annex 6

United States Patent Application 20160222840 Kind Code A1
Vaseleniuck; Darrick ; et al. August 4, 2016

MODULAR ROTARY VALVE APPARATUS

Abstract

A modular **rotary** valve apparatus includes: a plurality of separate valve barrels coupled to each other and arranged end-to-end along an axis so as to define a valve shaft, each valve barrel having an annular peripheral surface extending between forward and aft end faces, and an aperture extending transversely therethrough communicating with the peripheral surface on opposite sides.

Annex 7

United States Patent Application 20160160751 Kind Code A1
Koch; RandyJune 9, 2016

Rotary Internal ***Combustion Engine***

Abstract

A **rotary** internal **combustion engine** includes an arcuate compression chamber, an arcuate expansion chamber, an output shaft, and a piston coupled to the output shaft for movement through the arcuate compression chamber and the arcuate expansion chamber. The piston has a leading end, a trailing end, an inlet valve that is located at the leading end of the piston for receiving a compressible fluid from the compression chamber and an outlet valve that is located at the trailing end of the piston for expelling a **combustion** gas into the arcuate expansion chamber.

Annex 8

United States Patent Application 20160273446 Kind Code A1 LARSON; GERALD L. September 22, 2016.

*ROTARY **ENGINE***

Abstract

A **rotary engine** has a circular shaped rotor and stator. The rotor has an expansion chamber for **combustion** and the stator has thrust director gate that passes through a passageway through the stator face to alternatively ride along the face of the rotor or to ride within the expansion chamber of the rotor. Passageways through the stator and radially close to the thrust director gate allow for air, fuel, and an ignition source to pass through the stator allow for ignition and expansion of the fuel within the expansion chamber. The thrust director allows for thrust to be applied to the rotor to rotate the rotor. An exhaust manifold, also passing through stator, and also radially aligned with the expansion chamber allows exhaust gases from **combustion** to be released from the expansion chamber as the rotor rotates the expansion chamber adjacent to the exhaust manifold.

Annex 9

United States Patent Application 20150101557 Kind Code A1
Reisser; Heinz-Gustav A. April 16, 2015

ROTARY PISTON INTERNAL ***COMBUSTION ENGINE***

Abstract

Aninternal**combustion engine,**andmore particularly a **rotary** internal **combustion engine, is provided with said engine** having multiple **combustion** chambers delimited by piston heads and an **engine** housing wall that defines at least a section of a torus. Additionally, a method for operating the internal **combustion engine** is described.

Printed by Books on Demand GmbH, Norderstedt / Germany